I0503508

Table of Contents

Figure Table

EXECUTIVE SUMMARY

A biometric is a measurable physical characteristic or personal behavior trait used to recognize the identity or verify the claimed identity of an individual. Fingerprints are an example of a physical biometric characteristic. Behavioral biometric characteristics like handwriting are learned and acquired over time.

Biometrics is the process of recognizing an individual based on measurable anatomical, physiological and behavioral characteristics. Employing biometrics can help positively identify adversaries, allies and neutral persons. This is particularly useful when facing adversaries who rely on anonymity to operate.

Biometrics is not forensics even though the two can, and often are, employed in concert. Forensics involves the use of scientific analysis to link people, places, things and events while biometrics involves the use of automated processes to identify people based on their personal traits. Because of the interrelationship between biometrics and forensics, the Department of Defense (DOD) intends to develop a single concept of operation (CONOP) in the future describing how biometrics and forensics can be employed in a complementary manner.

This CONOP describes how DOD employs biometrics across the full range of military operations. It applies to all DOD organizations. This CONOP does not address the use of biometrics in DOD business functions not related to military operations.

The DOD biometric process relies on five biometric actions and three analytical/operational actions:

1) Collect: Obtain biometric and related contextual data from an object, system, or individual with, or without, his knowledge.

2) Normalize: Create a standardized, high-quality biometric file consisting of a biometric sample and contextual data.

3) Match: Determine whether biometric samples come from the same human source based on their level of similarity.

4) Store: Maintain biometric files to make available standardized, current biometric information of individuals when and where required. Biometric files are initially enrolled and then subsequently updated as part of storing.

5) Share: Exchange standardized biometric files and match results among approved DOD, interagency and international partners in accordance with applicable laws, policies, authorities and agreements.

6) Analyze: To deliberately consider biometric and non-biometric information on an individual and reach logical conclusions. These conclusions can include his intent, affiliation(s), activities, location and behavioral patterns.

7) Provide: Exchange analysis and associated information on individuals among approved DOD, interagency and international partners in accordance with applicable laws, policies, authorities and agreements.

8) Decide/Act: Take action based on a biometric file's match results and analysis of associated information.

By logically sequencing the biometric actions into the biometric process (Figure 1), the DOD can positively identify people and take appropriate action. This CONOP provides vignettes to illustrate this.

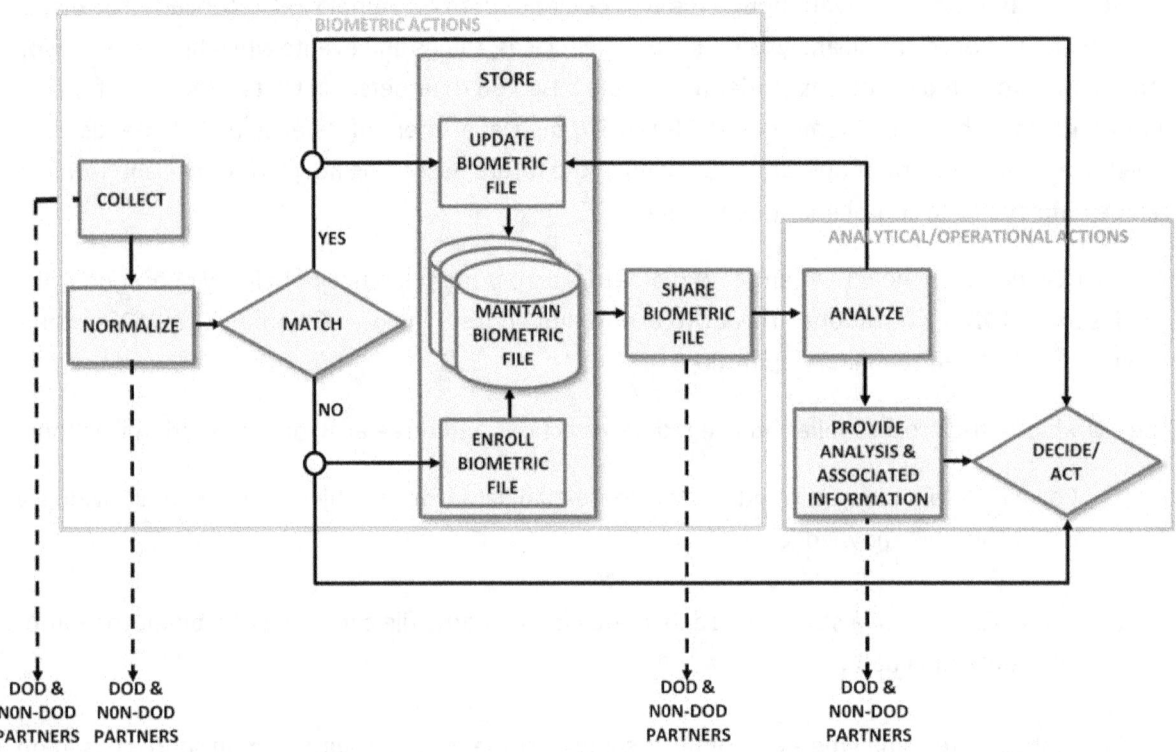

Figure 1: Biometric Process

(Note: Solid lines represent exchanges of information that must occur to support the decide/act action. Dashed lines represent additional exchanges of information to other entities.)

Biometrics operations are inherently legally intensive. Throughout the biometrics process, commanders and leaders must ensure operations are conducted in accordance with applicable laws, policies, authorities and agreements.

Whereas the biometric process progresses from collection of biometric samples to decision/action on them, the biometric cycle for military operations provides a method for integrating biometrics into military operations. It relies on six related activities as depicted and explained below:

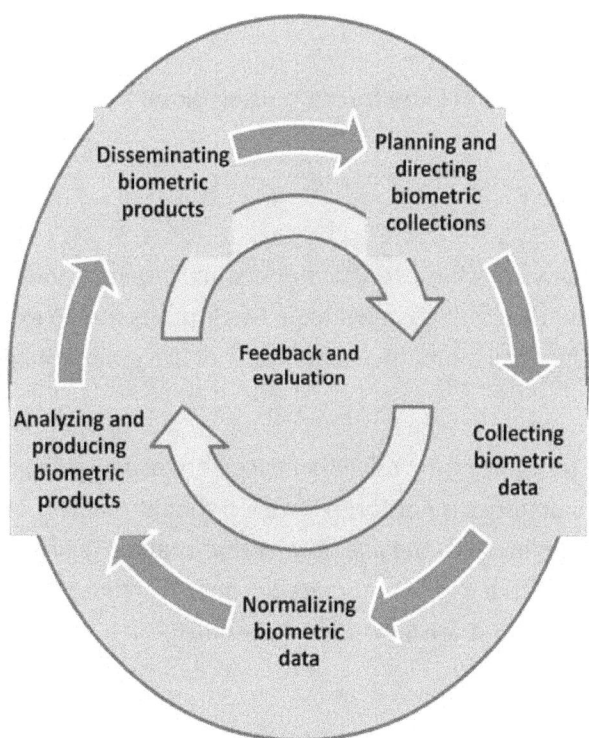

Figure 2: DOD Biometric Cycle for Military Operations

When planning and directing biometric collections, commanders identify which information requirements can be met with biometrics; determine what biometric data is needed; analyze where, when, how and by whom the data should be collected; what resources are required for collection; what biometric products are to be produced; and how those products will be disseminated.

Units collect and transmit biometric samples and contextual data in accordance with authoritative direction and their estimate of the situation. It is worth noting that collection of biometric data can be done by non-DOD authorities, a partner nation or agency for example, and then shared with the US military.

Normalizing biometric data converts it into forms that can be readily used by DOD. For example, if biometric files were obtained through a sharing agreement with another country, the contextual data may need translating prior to being input into DOD automated systems.

During analysis and production, biometric information, contextual data, and other related information and intelligence is integrated, evaluated, analyzed and interpreted to create finished biometric products that meet the commander's information requirements.

During dissemination, biometric products are delivered to and used by tactical units or other organizations.

During evaluation and feedback, units or organizations assess the effectiveness of biometrics efforts and adjust as needed.

1.0 PURPOSE

This CONOP's purpose is to assist US military forces in using biometrics across the full range of military operations.

2.0 DEFINITIONS

A biometric is a measurable physical characteristic or personal behavior trait used to recognize the identity or verify the claimed identity of an individual. Fingerprints are an example of a physical biometric characteristic. Behavioral biometric characteristics like handwriting are learned and acquired over time.

Biometrics is the process of recognizing an individual based on measurable anatomical, physiological and behavioral characteristics. Biometrics is not forensics even though the two can, and often are, employed in concert. Forensics involves the use of scientific analysis to link people, places, things and events while biometrics involves the use of automated processes to identify people based on their personal traits. Additional definitions are located at Appendix C.

3.0 SCOPE AND APPLICABILITY

This CONOP applies to DOD organizations employing biometrics in military operations while under DOD control. It is not applicable to DOD business functions unrelated to military operations such as enrollment as a military dependent. Because of this CONOP's broad nature, it does not reference specific biometrics equipment or systems.

4.0 OPERATIONAL ENVIRONMENT

The "Capstone Concept for Joint Operations" (CCJO) of 15 Jan 2009 describes the expected joint operational environment as dynamic and unpredictable. As described in the CCJO, the future operating environment will be characterized by uncertainty, complexity, rapid change, and persistent conflict.

Some adversaries will continue to rely on relative anonymity amongst the world's population to harm the US and its allies. Without being positively identified, enemies can remain hidden while pursuing their nefarious aims. The effective use of biometrics will help remove our enemies' critical capability of anonymity and expose them to counter action.

5.0 APPLYING BIOMETRICS

Biometric operations are undertaken to remove adversary anonymity and positively identify other people. In some circumstances, identifying other people can include US military personnel. For instance, biometrics may be used by recovery forces to positively identify isolated US military personnel or biometrics may be incorporated in sensitive site exploitations to help determine the whereabouts of missing US military personnel.

The DOD conducts biometric operations globally and across the range of military operations. This includes rule of law operations and security cooperation activities undertaken to favorably shape the environment.

US military forces conduct biometric operations independently as well as "by, with and through" other countries and in cooperation with other US government agencies. Typically this is done through mutually agreed upon exchanges of information. These agreements often contain caveats as to how and when information may be stored, shared and what it can be used for.

Combatant commanders may conduct biometric operations within their geographic or functional areas of responsibility. They allocate resources, organize and task their forces based on their missions and estimate of the situation. In addition, combatant commanders engage interagency and international partners and other combatant commands to conduct biometric operations. Since each combatant command's area of responsibility is different, the legal frameworks in which they conduct biometric operations are different.

DOD may enter into information sharing agreements with other countries to support combatant commander biometric efforts. Combatant commanders make foreign disclosure decisions involving biometric data on non-US persons in the absence of agreements and build partner nation biometric capacity through security cooperation and intelligence cooperation activities.

Since many countries collect biometrics more for law enforcement, border security, population management and immigration than military purposes, combatant commanders may consider engaging other countries' non-military government agencies through appropriate channels to facilitate biometrics sharing on non-US persons. For instance, a combatant commander may engage through a US embassy's country team.

Biometrics can be employed in all phases of a campaign or major operation. Annex A provides vignettes to illustrate this. Biometrics can:

- Shape events favorably toward US interests by assisting foreign governments and US law enforcement agencies in their efforts to maintain a secure environment and protecting US military forces
- Deter or dissuade an adversary who relies on anonymity as a critical capability
- Assist a joint force in seizing the initiative and dominating the situation
- Stabilize a situation by supporting civil affairs
- Enable civil authorities in maintaining, or establishing, rule of law and other governance functions

On the tactical level Commanders may use biometrics in: human terrain mapping, census taking, access control, personnel screening, incident response, personnel recovery, identification of human remains, detainee management and sensitive site exploitation among other things.

5.1 DOD BIOMETRIC PROCESS

Biometrics operations are inherently legally intensive. Throughout the biometrics process, commanders and leaders must ensure operations are conducted in accordance with applicable laws, policies, authorities and agreements. The DOD biometric process relies on five biometric actions and three analytical/operational actions that lead from the collection of biometric data to decision/action on it. The actions are:

1) Collect: Obtain biometric and related contextual data from an object, system, or individual with, or without, his knowledge.

2) Normalize: Create a standardized, high-quality biometric file consisting of a biometric sample and contextual data.

3) Match: Determine whether biometric samples come from the same human source based on their level of similarity.

4) Store: Maintain biometric files to make available standardized, current biometric information of individuals when and where required. Biometric files are initially enrolled and then subsequently updated as part of storing.

5) Share: Exchange standardized biometric files and match results among approved DOD, interagency and international partners in accordance with applicable laws, policies, authorities and agreements.

6) Analyze: To deliberately consider biometric and non-biometric information on an individual and reach logical conclusions. These conclusions can include his intent, affiliation(s), activities, location and behavioral patterns.

7) Provide: Exchange analysis and associated information on individuals among approved DOD, interagency and international partners in accordance with applicable laws, policies, authorities and agreements.

8) Decide/Act: Take action based on a biometric file's match results and analysis of associated information.

Logically sequencing the actions produces the DOD biometric process illustrated in Figure 1.

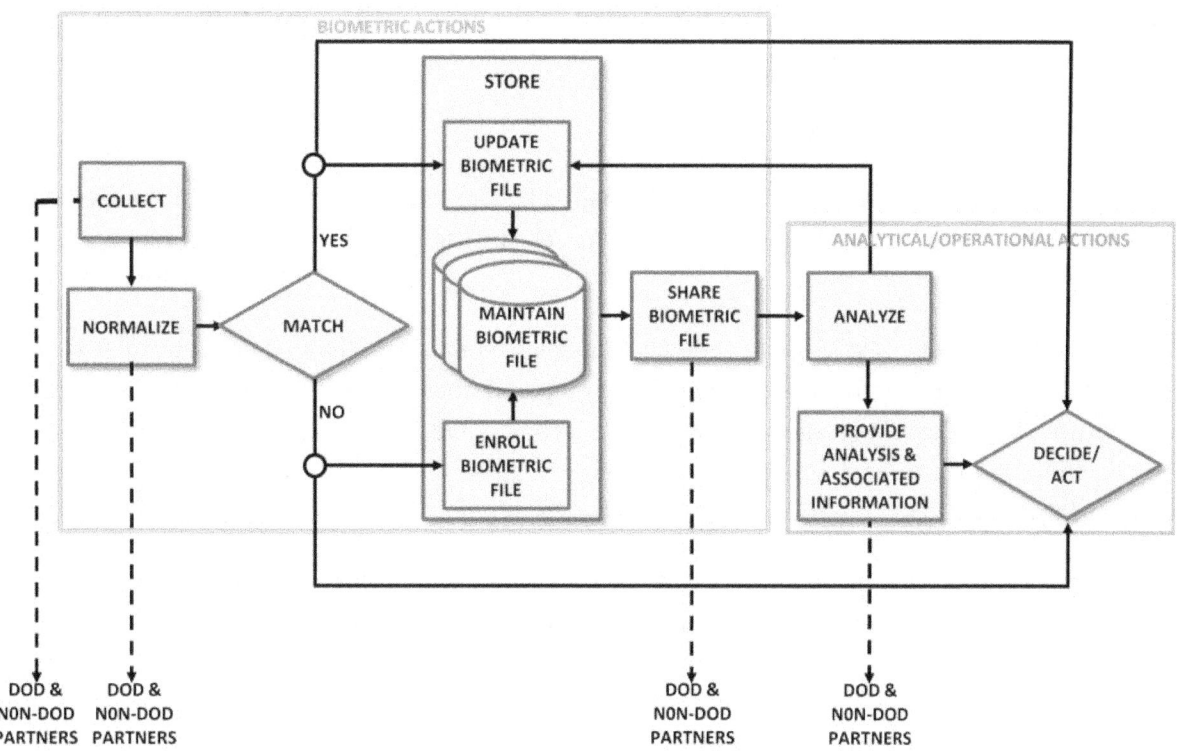

Figure 1: Biometric Process

(Note: Dashed lines represent additional exchanges of information to other entities.)

5.2 DISCUSSION OF DOD BIOMETRIC AND ANALYTICAL/OPERATIONAL ACTIONS

5.2.1 COLLECT BIOMETRIC SAMPLE

Collection is the obtaining of a biometric sample and related contextual data from an object, system, or individual with, or without, his knowledge.

It is important to note that collection of biometric samples can be done by other countries and non-DOD US government or non-government agencies and then, when appropriate and permissible, shared with the DOD. Commanders should obtain appropriate level legal review prior to collecting biometric data to ensure compliance with law, policy, and obligations with foreign agreement. Biometric collection operations should also be coordinated with other technical experts as dictated by the circumstances to provide the best results. These technical experts may include forensic technical examiners, network architects and communications specialists. Pre-planning before collection begins is important to ensuring the usability of collected data.

Collection can be done with, or without, the individual's knowledge. For DOD use, the biometric sample can be a fingerprint, facial image, palm print, iris image, handwriting, or voice sample. Finger and palm prints may be latent prints, e.g. taken from a surface touched by the individual using forensic methods vice taken directly from the individual. Contextual data includes situational information associated with a collection event. For instance, where and when the biometric sample was collected; under what circumstances it was collected; why it was collected; the claimed identity of the individual; and biographic data is usually collected.

5.2.2 NORMALIZE BIOMETRIC SAMPLE

Normalization is the process of transforming biometric files so they are in a standard format and meet a specified level of quality. This ensures biometric files can be used by DOD and other automated biometric systems.

DOD biometric collection systems are generally capable of creating normalized biometric files at the point of collection. Biometric files obtained from other countries however are likely to have contextual data written in a foreign language and use formats incompatible with DOD systems. Incompatible files such as these require normalization for subsequent matching, storing and sharing by DOD. This can be painstaking and resource intensive, but a critical task.

Once a biometric file is normalized, it may be transmitted to a data source for matching if the biometric collection system does not allow for matching at the point of collection.

5.2.3 MATCH

Matching is the process of deciding whether, or not, a biometric sample and a stored template come from the same human source based on their high level of similarity. Matching consists of either a one-to-one (verification) or a one-to-many (identification) search.

- Verification: Verification is a task where a biometric system attempts to confirm an individual's claimed identity by comparing a newly submitted sample to already enrolled samples. It answers the question, "Is this person who they say they are?"
- Identification: Identification is a task where a biometric system searches a database for a reference matching a submitted biometric sample and, if found, returns a corresponding identity. Identification answers the questions, "Have we ever collected biometric information from this individual before and, if so, what was their claimed identity?" Identification is "closed-set" if the person is known to exist in the database. In "open-set" identification (sometimes referred to as "watch listing") the person is not guaranteed to exist in the database. The biometric system must determine where the person is in the database, and then return his identity.

Once matching is complete, the collected biometric sample and contextual data can be enrolled into the repository as a new file, used to update an existing file or deleted entirely.

5.2.4 STORE

During storage, DOD users enroll or update and maintain biometric files at one of three types of source locations. The purpose is to make available standardized, comprehensive and current biometric information on individuals. The vignettes in this CONOP illustrate applications of stored biometric files in operations.

As part of storage, the biometric files may be marked and assigned caveats and have permissions set for appropriate access and use.

For the purpose of this CONOP, a source is a database and infrastructure that stores biometric files. Storage requirements can be as complex as a data mega-center or as simple as a hand held device. There are three types of biometric storage sources:

- Authoritative source: An authoritative source is the primary DOD approved repository of biometric information. It provides a strategic capability for access to standardized, comprehensive and current biometric files within the DOD and for the sharing of biometric files with joint, interagency and designated multinational partners. The DOD may designate more than one authoritative source for various populations in accordance with applicable law, policy and directives. All DOD operational applications should be designed to acquire biometric files from the appropriate authoritative source.
- Local trusted source: A local trusted source is a sub-set of the authoritative source that is established to accomplish a specific function within an operational mission. Reasons for establishing a local trusted source include: insufficient network connectivity able to provide adequate access to the authoritative source or an operational need for closed-loop access or permission application.
- Local un-trusted source: A local un-trusted source is a local repository of biometric files that have not been enrolled with an authoritative or local trusted source. In many cases, local un-trusted sources are established for missions of short duration or to satisfy political, policy, or legal restrictions related to the sharing of biometric information.

5.2.5 SHARE

Sharing is the exchange of standardized biometric files or match results.

Sharing of biometric files is conducted through DOD-approved biometric information exchange portals and other sharing systems among DOD, interagency and multinational partners as mutually agreed upon or otherwise allowed by law and policy. As stated before, biometric sharing agreements with partner nations may have caveats as to how, when and with whom biometric information they provide may be shared. Commanders should obtain appropriate level legal review prior to sharing biometric files or match results to ensure compliance with law, policy, and foreign agreement obligations.

5.2.6 ANALYZE

During analysis we integrate biometric, associated, and intelligence information about a person and determine what it all means. We query information repositories for non-biometric information associated with an identity to answer the question, "Do we know anything else about this person in addition to his biometric file like his location, personal associations, or activities?" If there is associated information, it may be linked to a biometric file and this link can be used to prompt users of the associated information.

5.2.7 PROVIDE

In providing we disseminate information, analysis and conclusions to appropriate organizations and entities in a timely manner.

5.2.8 DECIDE/ACT

The decide/act action is the response (either automated or human directed) to the results of the analysis and/or match.

5.3 DOD BIOMETRIC CYCLE FOR MILITARY OPERATIONS

Whereas the biometric process progresses from collection of biometric samples to decision/action on individuals, the biometric cycle for military operations provides a method to integrate biometrics into military operations. The two are closely related, mutually dependent, and share some activities.

The biometric cycle for military operations relies on six related activities as depicted and explained below:

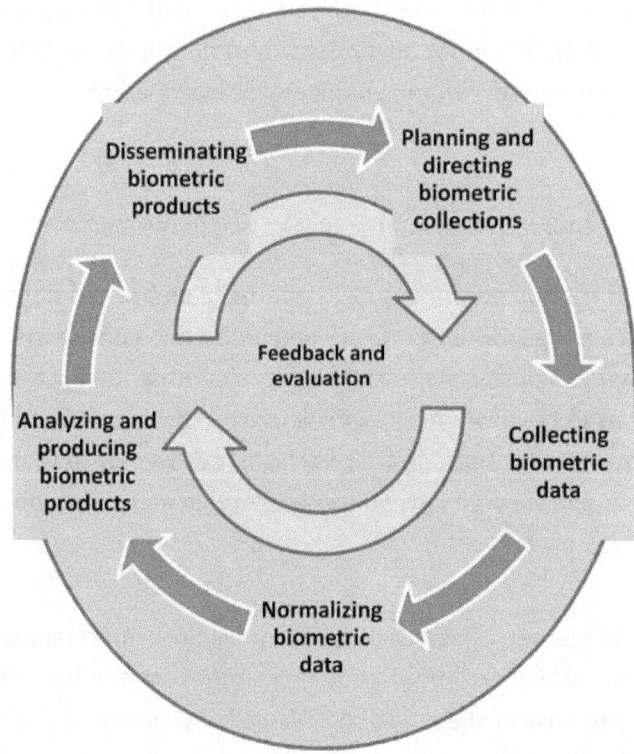

Figure 2: DOD Biometric Cycle for Military Operations

5.3.1 PLANNING AND DIRECTING BIOMETRIC COLLECTIONS

When planning and directing biometric collections, commanders identify which information requirements can be met with biometrics; determine what biometric data is needed; analyze where and when the data should be collected; what resources are required for collection; what biometric products are to be produced; and how those products will be disseminated.

Commanders and their staffs should also consider the utility of biometrics in recovering captive or isolated US or friendly personnel. Specifically, biometrics can be used to prove the captured or isolated people are still alive and positively identify them provided their biometric data has already been collected and is available to recovery forces.

5.3.2 COLLECTING BIOMETRIC DATA

Units collect and transmit biometric samples and contextual data in accordance with authoritative direction and their estimate of the situation. It is worth noting that collection of biometric data can be done by non-DOD authorities, a partner nation or agency for example, and then shared with the US military.

5.3.3 NORMALIZING BIOMETRIC DATA

Normalizing biometric data converts it into forms that can be readily used. For example, if a large number of biometric files were obtained subsequent to an agreement with another country, the contextual data may need translating or the files may need to be normalized prior to being input into DOD automated systems.

5.3.4 ANALYZING AND PRODUCING BIOMETRIC PRODUCTS

During analysis and production, biometric information and contextual data is integrated, evaluated, analyzed and interpreted to create finished biometric products that meet the commander's information requirements or some other need. For instance, biometric products may be used to help build a prosecutable case for a foreign government during rule of law operations.

5.3.5 DISSEMINATING BIOMETRIC PRODUCTS

During dissemination, biometric products are delivered to and used by tactical units or organizations supporting them.

5.3.6 FEEDBACK AND EVALUATION

During evaluation and feedback, units assess the effectiveness of biometric efforts and adjust as needed. Evaluation and feedback may also serve to refine biometric collection requirements and priorities in phased operations as the combination and sequencing of offensive, defensive and stability activities is adjusted.

6.0 RISKS AND MITIGATION

6.1 PROTECTION OF BIOMETRIC DATA

Biometric data can be misused. DOD must protect and secure its biometric data to ensure its integrity and security. During system development, business operations, and execution of operations collected biometric data must be protected in a distributed network environment. This protection ensures the credibility and integrity of the biometric sources, files, and authoritative databases.

6.2 COUNTERMEASURES

Our enemies are adaptive and continually seek new ways to counter our biometrics technologies and processes. Innovative research, engineering and testing are necessary to ensure that we incorporate technological advancements into our future biometric capabilities. In order to maintain our technological advantage we must ensure requisite analytic support; joint experimentation of concepts and pilots; and necessary standards continue to be developed and resourced. The DOD must also use current information assurance practices in order to protect against unlawful access or inadvertent release of biometric information.

6.3 RESTRICTIONS

Domestic and foreign laws, international agreements, policies, regulations and socio-cultural inhibitions may prohibit or restrict the employment of biometric capabilities. Biometric systems must be flexible to accommodate restrictions and changes to an individual's status as they occur. The DOD may cooperate with domestic and foreign partners to employ biometrics capabilities while respecting agreements, laws, policies, standards, regulations and department direction. Because DOD's biometric systems are largely automated, they must be adaptive to adhere to differing agreements and restrictions while processing a large volume of transactions without reliance on human oversight of each one.

6.4 DATA QUALITY AND LATENCY

We must use standardized, high quality files that are properly transmitted and loaded promptly in order to most effectively use biometrics. A poor quality file may preclude it from being matched. A file that is not promptly transmitted and uploaded into the appropriate data source(s) reduces its utility. Standardization of biometric collection devices and normalization of files, together with matching, storing and sharing standards are critical in order to share files with partner organizations. Once standards are established, they must be adhered to for DOD's biometric systems to work properly in supporting military operations.

7.0 IMPLICATIONS

Employment of biometrics has implications across the areas of doctrine, organization, training, materiel, leadership and education, personnel, facilities, policy, standards, data sharing and research and development. These implications must be addressed in related documents. Potential implications include:

- The DOD must continue to influence, establish, direct, adopt, support and enforce complementary biometric related policies and standards, both within DOD and with external partners.

- Applying biometric capabilities requires cross-domain solutions that are interoperable with US and coalition processes and systems. Consequently, the DOD must continue to influence, establish, adopt, support and enforce national and international biometric standards and operating rules to ensure the requisite interoperability is achieved. Additional detail must be developed in implementing documents.

- Formal biometric data sharing policy guidance must be established with joint, interagency (e.g., Federal Bureau of Investigation criminal justice information systems, Department of Homeland Security visitor and immigration systems), state, local, tribal and international partners to fully exploit the capability described in this CONOP.

- Application of DOD's biometrics capabilities must adhere to US law, policy, and agreement. Legal review of planned or ongoing activities is required in order to assure this.

- Successful employment of biometrics relies on several information assurance (IA) capabilities, including confidentiality, integrity, availability and non-repudiation. Biometric and biometric-enabled systems must adhere to the data security requirements specified by IA related directives.

- Because sharing biometric files with joint, interagency and multinational partners may be necessary, the DOD intends to keep the classification of biometric files at the lowest level acceptable to the mission and conditions of collection. If special circumstances make it necessary to classify a biometric file, the file will require processes established in accordance with applicable laws, policies, directives and guidance from national and OSD governance bodies.

- Employing biometrics can create significant demands on communication systems. Communication systems must be able to support operational timeliness requirements for biometrics. Meanwhile, reducing bandwidth requirements through data compression and templating should be pursued.

- Research and development activities must support and involve interagency participation.

- US national and DOD biometrics organizations must identify and resource authoritative sources of associated information and analysis of individuals.

- DOD must integrate multimodal biometric systems for collection, matching and storage.

- Matching is a statistical process. Consequently, there is a possibility for error, such as a false match or a false non-match. This possibility must be considered when developing standards, procedures and making decisions.

A. VIGNETTES

Military operations are conducted across the conflict continuum. The following vignettes describe how the DOD can employ biometrics across the range of military operations as depicted in Figure 3. The vignettes describe situations where establishing an individual's identity through biometrics yields an advantage. Some capabilities described in the vignettes are assumed to be future capabilities. For each vignette, assume the described operations have been reviewed and found consistent with all applicable foreign and domestic laws, regulations, policies and agreements.

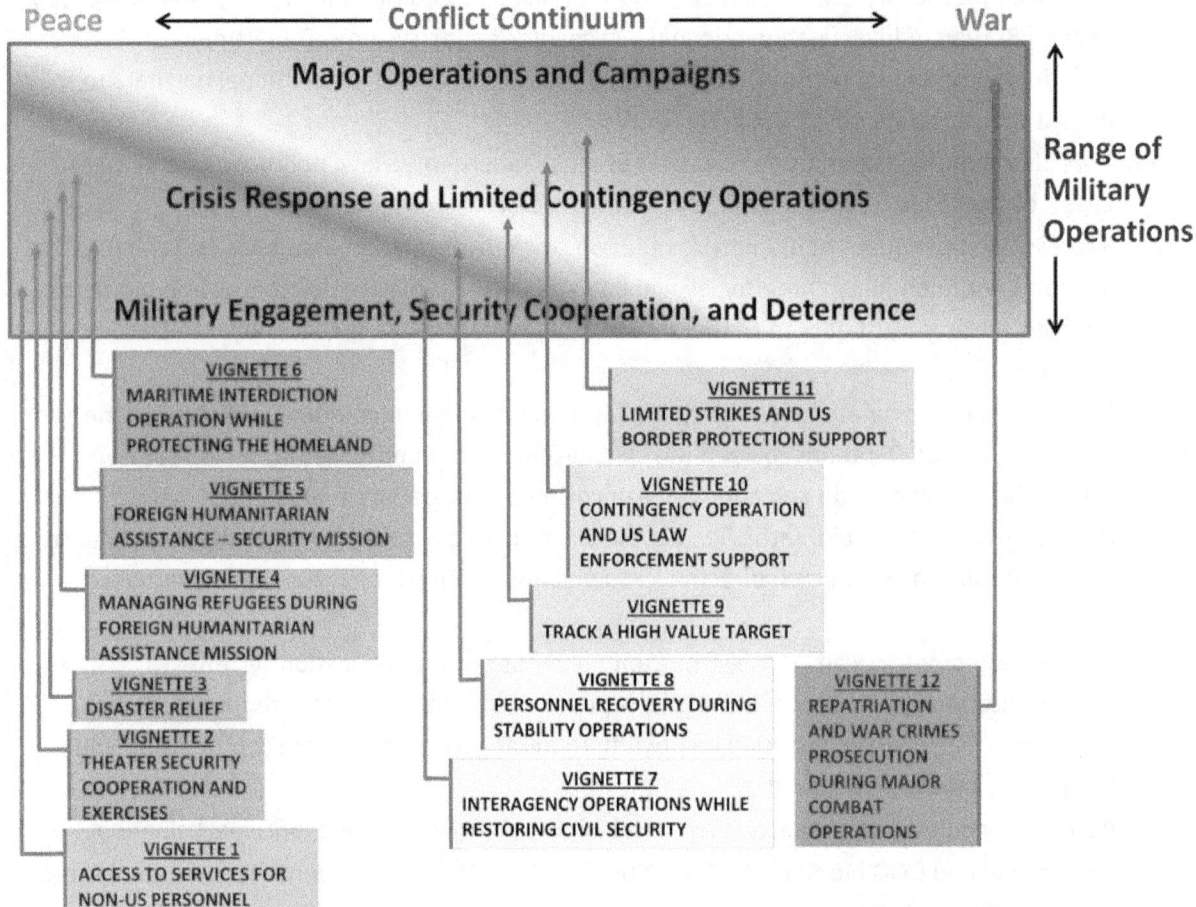

Figure 3: Biometrics Across the Range of Military Operations

A.1 ACCESS TO SERVICES FOR NON-US PERSONNEL

While operating in the host nation, the US contracts with local nationals to provide labor and services. As a condition of employment, the laborer must provide individual identity information and biometric samples for screening and background check purposes. Biometric samples are taken and matched against both host nation and US authoritative sources. Both positive and negative matches result in the update and enrollment of individual biometric files, respectively. Additionally, once stored, these biometric files are shared with host nation and US government non-DOD parties for subsequent analysis and fusion of applicable biometric and associated information (e.g., criminal records). Based on this exhaustive research, the US military decides whether to offer employment and issue biometrics-enabled identity cards to the job applicant. Successfully screened laborers receive identity cards that they must display to access the base and receive wages for work performed. Biometric matching of all laborers is conducted on payday to confirm identity prior to payment.

One individual has lost his ID card, but his biometric sample matches his file in the local trusted source. Analysis of associated information by finance personnel indicates that he worked every day. He is paid and his biometric file is updated.

A second individual presents his ID card and a biometric sample. He is matched to the local trusted source and identified as having been fired two days ago. Finance personnel determine how much he is owed and he is required to surrender his ID card (he had claimed to have lost it the day he was fired) upon leaving the installation.

A third individual provides his ID card. His picture appears to match; however, his biometric sample does not match to any individual stored in the local trusted source. On-scene analysis reveals he is the brother of an actual worker. The individual is detained and escorted off base. A biometric file is created and stored at the local trusted source and later shared, along with other non-biometric information, with non-DOD partners. Additionally, a prompt with this information is attached to his biometric file for possible disciplinary action the next time he attempts to enter the base. The brother's (actual worker) biometric file is also identified to indicate his credentials have been compromised and this information is provided to other relevant authorities.

Tasks achieved utilizing biometrics:

- Identify unknown individuals during tactical operations
- Manage local populations during military operations
- Enable information assurance (authorize access to privileges)
- Control physical access

A.2 THEATER SECURITY COOPERATION AND EXERCISES

The US military furthers security cooperation through medical civic action programs (MEDCAPs) to remote regions of allied nations in conjunction with port visits and combined military exercises.

During an annual combined exercise, an Army medical detachment executes MEDCAPs in a number of villages within the exercise AOR. Army medics collect biometric information on those who receive vaccinations and medical treatment during the MEDCAPs. The biometric information is collected with consent from the patients and the host nation government. Biometric files are enrolled and stored for each individual receiving treatment and/or vaccinations. These biometric files are linked to subsequent treatment and vaccination records stored in other repositories of associated information.

The following year a different Army medical detachment deploys to the AOR to perform MEDCAPs. At the first village, Army medics encounter far more villagers awaiting vaccination than anticipated, creating concern that the amount of on-hand vaccine is insufficient. To assist the ongoing mission, a repository of associated information has been established during previous MEDCAP exercises. Biometric samples are collected on each person awaiting vaccination and matched to the local-trusted source. Numerous positive matches occur. These match results are compared against the repository of associated information to identify which individuals received vaccinations in the past. Analysis of the match results and the repository of associated information reveals that a large number of those awaiting vaccination have already received the vaccine during previous MEDCAPs and do not require an additional dose.

Relying on the biometric data, the on-scene commander orders vaccination of only those with no biometric match and those with biometric matches whose linked medical treatment record does not indicate the vaccine was previously received. The villagers are briefed accordingly.

The Army medics successfully complete the MEDCAP with the vaccine on hand. The on-scene commander is confident that the total supply of vaccination is sufficient for future MEDCAPs based on the biometric matches experienced in this initial MEDCAP.

Tasks achieved utilizing biometrics:

- Manage local populations during military operations
- Enable information assurance (authorize access to privileges)
- Share identity information

A.3 DISASTER RELIEF

The US Government is responding to a request from a country that has experienced a catastrophic disaster. The disaster has created the immediate need to locate, rescue and manage the affected population.

The host government approves the multinational response force to collect biometric samples from the civilian population to assist with disaster relief efforts with the stipulations that: (a) the biometric information only be used to identify individuals located and rescued and to manage the flow of casualties and the displaced population; and (b) the biometric information not be removed from the country.

Biometric data is collected as the affected individuals are rescued, treated or entered into the refugee management process. Because of the scale of the disaster, much of the biometric data from affected people is collected by the host government and then given to the DOD. After normalizing the data, DOD personnel store the collected biometric files on a local un-trusted source and use them as the reference set against which subsequent matches are made. As personnel are placed aboard transportation, provided medical care or basic services at a disaster relief site, the individuals' biometrics are the "tokens" that authorize their access. In each instance, once the biometric file is matched, the identity is referenced against repositories of non-biometric information such as camp rosters, medical records, records of service provided, transportation logs, etc. to enable better management of services provided and needs of the population. This data and the collected biometrics are shared with the host nation and coalition partners to assist in integrating their relief efforts with those of US forces. The host nation also compares the collected information with repositories of non-biometric data that have survived the disaster (tax records, census data, etc) to assist in the speedy location and reunion of families. At the request of relief organizations, the national government shares the biometric data and identification results with NGOs and neighboring countries affected by the refugee flow.

Tasks achieved using biometrics:

- Manage local populations during military operations
- Manage emergency situations
- Share identity information

A.4 MANAGING REFUGEES DURING FOREIGN HUMANITARIAN ASSISTANCE MISSION

The US military is responding as part of an international disaster relief effort. Thousands of injured are being treated and awaiting further treatment as soon as field medical hospitals are assembled and operational. All individuals who receive medical attention within the disaster area are immediately enrolled in a biometric local un-trusted source that has been established for management of the refugees. All treatment records are linked to their respective biometric files. Many of the injured, after being initially treated, voluntarily relocate within the disaster area. This movement is making it difficult for medical personnel to efficiently provide medical services or track patients for follow-up treatment.

Navy corpsmen are performing triage for refugees arriving by buses at one of the newly established US field hospitals. The corpsmen collect biometric samples from each refugee for identification purposes as part of the initial medical assessment process. The biometric files are then sent for matching against the local un-trusted source to assist with the identification of the individual and retrieve any available treatment history.

A refugee who cannot be matched against the local un-trusted source is enrolled as a new biometric file. All subsequent medical treatment will later be linked to the file of that individual. When a refugee is positively matched against the local un-trusted source, links to his medical history are accessed and his prior treatment records are retrieved. Subsequent treatment is updated in the refugee's medical record so that information can be accessed by others again in the future through utilizing the established net-centric links between the non-biometric repository (medical files) and his biometric file. The corpsman uses these medical records to aid in triage.

Tasks achieved utilizing biometrics:

- Manage local populations during military operations
- Manage emergency situations
- Share identity information

A.5 FOREIGN HUMANITARIAN ASSISTANCE—SECURITY MISSION

The US and multinational partners operate from several dozen military bases in an allied nation and contract locally for a wide range of services, such as vehicle rental and maintenance, civil construction, provisioning of food and water and waste removal. Biometrics are collected to support a wide range of activities from base access to monitoring of all contracting activities. All biometric data are matched against the local-trusted source and repositories of associated information for the purposes of vetting. All samples reveal a negative match and are enrolled in the local-trusted source and transmitted to the authoritative source.

Several base contracting officers encounter a dishonest local contractor who is awarded contracts and receives partial payment but never performs the work – essentially disappearing with the money. This associated information is analyzed with relevant biometric data. This analysis is transmitted to the authoritative source, the individual's biometric file is identified and repositories of associated information are modified for future analysis to indicate he is barred from further contracts. This information is then shared with local-trusted sources and other interested parties.

The dishonest local contractor relocates to another region and applies for new US and coalition contracts using a different company name and false personal data. The contracting official collects his biometric sample and requests a match from the local-trusted source. A subsequent positive match reveals a prompt directing the user to relevant associated information indicating his previous activities and status. His bids are eliminated. The dishonest contractor's biometric file is updated with the newly collected biometric sample and contextual data and the attempt is shared with all appropriate authorities.

A newly-arrived disbursing officer is ordered into the local community to pay a contractor for recently completed work. This officer has never met the local national to whom he is to pay a large sum of cash. Following the directions provided by a local interpreter, the disbursing officer arrives at what he believes is the office of the intended contractor. Unbeknownst to the disbursing officer, he has arrived at a fake contractor's office. As a condition of payment the supposed contractor provides his biometric information. A field match test reveals the presented biometric samples do not match the biometric file of the individual identified in the contract. The disbursing officer refuses to pay despite the local interpreter's and contractor's insistence.

Upon returning to base the disbursing officer provides the collected biometric information and his incident report to the Provost Marshal for investigation with the local police. The local interpreter is immediately detained on-base for questioning. The fraudulent contractor's biometric file is enrolled and stored within the local-trusted source, transmitted to the authoritative source and shared with interested parties. Upon completion of the investigation, the Provost Marshal concludes that the contractor is a fraud. US military contracting offices operating within the region as well as the host nation update their respective repositories with this information.

Tasks achieved utilizing biometrics:

- Track a person of interest
- Manage local populations during military operations
- Control physical access
- Enable information assurance (authorize access to privileges)
- Share identity information

A.6 MARITIME INTERDICTION OPERATION WHILE PROTECTING THE HOMELAND

A US Coast Guard Law Enforcement Detachment (LEDET) is aboard a US Navy ship and operating under US Navy tactical control while conducting a compliant maritime interdiction operation seeking terrorists. After obtaining flag state consent, the LEDET boards a large container ship and collects biometric samples from each crewman. The data is transmitted to a DOD authoritative source and is followed up with acknowledgment of receipt. The biometric data is compared against all stored files and shared with mission partners. A subsequent match is made on three of the crewmen. Furthermore, the matched files show a link to the National Counter-Terrorism Center (NCTC) terrorist watchlist. The authoritative source updates the applicable biometric files with newly collected biometric samples and contextual data. The LEDET is informed of the match result and watchlist status.

Further analysis of the biometric files and additional associated information indicates the three crewmen have travel patterns consistent with those of previously apprehended terrorists. Based on this information, the on-scene commander detains the three crew members pending further disposition.

The on-scene commander further requests, and is granted, flag state authorization to conduct a detailed search of the vessel. In the course of the search, 40 undocumented individuals are discovered in a cargo hold. They are determined to be attempting illegal entry into the US. Also during the search, documents related to the design of an improvised nuclear device are discovered and collected. Biometric samples are next collected from the undocumented individuals. Their biometric data is transmitted to the DOD authoritative source and compared to all stored files; however, no match is made. Each undocumented individual's biometric data is enrolled into a biometric file, linked to the weapons of mass destruction (WMD) information and stored for later use. The biometric files and related associated information are shared with the mission partners and entered into interagency systems, including the maritime domain awareness systems, the FBI criminal justice information systems and DHS visit and immigration systems. The on-scene commander informs the appropriate authority and, after receiving flag state and US Government authorization, takes the undocumented individuals into custody pending further disposition.

Tasks achieved using biometrics:

- Identify an unknown individual during tactical operations
- Locate a person of interest
- Track a person of interest
- Enable information assurance

- Share identity information

A.7 INTERAGENCY OPERATIONS WHILE RESTORING CIVIL SECURITY

US and multinational forces are supporting a foreign country's rebuilding process, which is being undermined by smuggling into the state. The US and multinational forces are working largely "by, with, and through" the host government's institutions. The host government has only allowed US forces to use collected biometric data within the host nation and required all collected data be shared with them. Therefore, all biometric operations are conducted using local un-trusted sources.

In accordance with standard operating procedures (SOPs), a truck driver provides biometric samples to the border police at a remote international border crossing supported by US military personnel. The biometric samples and contextual information are transmitted to the local un-trusted source and subsequently compared to locally stored biometric files. The truck driver's biometric data does not match any file at the local un-trusted source and a negative response is provided back to the border police. The truck driver also is checked against local and national criminal records. The border police review the match result, associated information and other available situational information and clear the truck driver to continue. The biometric file is enrolled and stored at the local un-trusted source, as well as shared with US forces, multinational partners and non-governmental organizations (NGOs) operating within the country.

Several months later, the host nation's national police, supported by a US Government agency, conduct a raid on a drug-smuggler's safe house and seize numerous documents and other evidence. Biometric samples are collected using forensic techniques and compared to the local un-trusted source. A match is made between the latent fingerprint samples collected during the raid and the truck driver's previously collected biometric file. An analysis of the raid, as well as additional associated information, is completed and the truck driver's non-biometric reference information is updated with these new samples, identified for future matches and shared with all local sources within the country.

Several days later, the truck driver attempts to cross at a different border checkpoint. He submits his individual identification and a biometric sample for verification. The sample is compared against the truck driver's biometric sample on file, which alerts the border police to the prompt stored at the local un-trusted source. The truck driver is detained for questioning and his biometric file is updated with the newly collected biometric sample and contextual data.

Tasks achieved using biometrics:

- Identify an individual during tactical operations
- Locate a person of interest
- Track a person of interest
- Manage local populations during military operations
- Control physical access
- Share identity information

A.8 PERSONNEL RECOVERY DURING STABILITY OPERATIONS

US government, DOD, multinational and NGO personnel are conducting stability operations in a country coping with insurgent activity. Several civilians have been abducted.

A US special operations forces (SOF) team receives information from intelligence sources concerning the location of a US civilian contractor who has been held by insurgents for nearly 30 days.

Prior to conducting a rescue operation, the SOF team downloads digital biometric files and associated information on the captive from the authoritative source in order to verify the individual's identity.

During the operation, the team detains seven individuals at the site and collects their biometric data. Using their tactical biometric device, the team immediately matches one sample to the individual the unit was sent to recover. The team also uses associated information obtained from the contractor's firm to verify the identity of the individual. Other individuals are not immediately matched and their biometric files are transmitted, enrolled and stored at the authoritative source. The authoritative source acknowledges receipt of biometric files.

The team initiates handling protocols for the rescued captive and detains the remaining individuals. At the repository, the files are processed and stored for future use.

Tasks achieved using biometrics:

- Identify an individual during tactical operations
- Locate a person of interest
- Identify friendly forces

A.9 TRACK A HIGH-VALUE TARGET

While on patrol, a squad of Marines detects an improvised explosive device (IED). Explosive ordnance disposal technicians render the device safe, a forensics team manages to collect latent fingerprints and DNA samples, and the IED components are sent to a forward forensic facility for more analysis.

The latent fingerprints are formatted into a standardized electronic file, compared to samples on file and stored locally. There is no match at the local-trusted source and the data is enrolled into a biometric file. Both the electronic fingerprint file and DNA samples are transmitted to their respective authoritative source for further comparison. Acknowledgement of receipt is transmitted back to the local source. Matching at the authoritative source does not yield a DNA match and the sample is stored for further comparison. The fingerprint and DNA samples are also shared with coalition partners, revealing a fingerprint match to a suspected bomb-maker. Based on this identification, the coalition partner provides a facial photograph of the suspected bomb maker as well as other information gained from intercepted signals and captured documents.

Analysis of the shared biometric samples, signals and documents indicates the suspected bomb maker's last reported location was outside the joint area of operations in a country providing sanctuary. This analysis, as well as the photograph provided by multinational partners, is sent to the DOD authoritative source to update the biometric file. An alert (prompt) containing pointers to information located in non-biometric reference data is disseminated to tactical users to facilitate future data comparisons on their local biometrics systems should they encounter the individual.

A series of raids on suspected insurgent locations provides more biometric samples that are matched to the suspected bomb maker. This match information, the biometric files and the associated information from the previous analysis that led to his being tied to the IED incidents are shared with interested parties for further analysis. Analysis of associated information indicates that the suspected bomb maker is moving within the area of responsibility and provides locations he will likely move to. Cameras are positioned around those locations and provide photographs that identify the suspected bomb maker using facial recognition. Once the suspected bomb maker is located, a tactical unit conducts a raid to apprehend him.

The raid force encounters six men at the site, all with authentic-looking identification in their possession. Pictures of the bomb-maker provided to the raid force are outdated and do not closely resemble any individual at the raid site but a field biometric test matches the suspected bomb-maker. Analysis of that biometric match result and associated information from the previous analysis that tied him to the IED incident enable the raid force leader to decide to detain that man. The other men are released after collecting their biometric samples and comparing them against available repositories to determine if they had been encountered previously. All collected samples and contextual information are updated in their respective biometric files and annotated to reflect that the raid force encountered them in the company of a known bomb-maker. Other information found at the scene is also collected by the raid force and subsequently stored in a repository of associated information for use in later analysis.

Tasks achieved using biometrics:

- Identify an unknown individual during tactical operations
- Locate a person of interest
- Track a person of interest
- Enable information assurance
- Share Identity Information

A.10 CONTINGENCY OPERATION AND UNITED STATES LAW ENFORCEMENT SUPPORT

A squad on a patrol is attacked by armed fighters in native garb. After the initial skirmish, the fighters surrender their arms and are detained by US military forces. A search of the subjects' possessions reveals falsified identification documents from Iraq, Afghanistan and Pakistan.

Biometric samples are collected from each of the detainees and are transmitted to a DOD authoritative source. The data is compared against all files within the authoritative source and a positive match is made on two of the individuals. Match results indicate these two subject's biometrics have been found at a location containing bomb-making materials in Yemen around the time of the USS Cole attack.

After updating and storing the subjects' new biometric files, the DOD shares all of the biometric samples and associated information with the FBI's and Department of Homeland Security's (DHS) biometric databases. After analysis of available biometric and associated information, the subjects are nominated and promoted by the NCTC as known or suspected terrorists. The subjects' biometric files are identified and linked to the NCTC's terrorist watch list at the DOD authoritative source, as well as entered into the FBI's known or suspected terrorist (KST) database.

Several months later, the detainees are released to a foreign government for adjudication and repatriation.

Several years later, a US police department responds to a trespassing complaint at a local water treatment plant, which services a large metropolitan area. Two subjects are apprehended and fingerprints are taken at the police department's primary booking station. The fingerprints are transmitted to the FBI's fingerprint database and matches are made against the previously shared biometric files collected from the military detainees. Because the fingerprints have been entered into the FBI's KST file, the FBI CJIS Division Intelligence Group immediately alerts the Terrorist Screening Center (TSC) of the encounter. Upon notification, the TSC advises the local Joint Terrorism Task Force to investigate whether the trespassing act was an indication of a terrorist threat to the nation.

Tasks achieved using biometrics:

- Identify an unknown individual during tactical operations
- Locate a person of interest
- Track a person of interest
- Enable information assurance
- Share identity information

A.11 LIMITED STRIKES AND UNITED STATES BORDER PROTECTION SUPPORT

Allied forces are supporting a foreign country's operation to neutralize a suspected WMD bomb-making facility within the country's borders. During a successful raid of the facility, US military forces locate stockpiles of IEDs and detain several subjects in connection with the operation.

The subjects are turned over to the foreign country's government after biometric samples and contextual data are collected and transmitted to a DOD authoritative source. The data is compared against all files within the authoritative source and no matches are made on any of the individuals. The DOD authoritative source enrolls the new biometric files. The DOD shares the biometric files and associated information with the FBI. There are no matches within the FBI's database.

Several months later, the subjects escape from the foreign government's prison system.

Several years later, DHS Customs and Border Protection (CBP) collects a visitor's fingerprints during a primary border-entry check. The CBP transmits the biometric information to the DHS authoritative source. Through system interoperability with the FBI's biometric database, DHS identifies one of the subjects as having been previously detained at the WMD bomb-making facility.

Upon notification of the match, the primary border check escalates to a secondary CBP inspection and an investigation into the encounter is initiated. After a more detailed inspection, an IED is found concealed in the subject's vehicle and is later determined to be a WMD. The subjects are immediately detained and handed over to the FBI for further questioning.

Tasks achieved using biometrics:

- Identify an unknown individual during tactical operations
- Track a person of interest
- Control physical access
- Share identity information

A.12 REPATRIATION AND WAR CRIMES PROSECUTION DURING MAJOR COMBAT OPERATIONS

US and allied forces have begun major combat operations after several months of intense diplomatic activity. They are acting under a UN resolution authorizing the use of force. Their purpose is to defeat a country's invasion of one of its neighbors.

The UN forces take a large number of prisoners of war early in the campaign. Facial photographs and fingerprints are taken from the prisoners in order to provide the opposing nation and appropriate international organizations an accounting of the identity of the prisoners, their health and location. The fingerprints and photographs are also enrolled into DOD's authoritative database of biometric samples.

Several mass graves and caches of looted valuables are discovered by UN forces as liberation of the country continues. Latent fingerprints are recovered from the sites using forensic techniques and matched to twenty enemy prisoners of war. When questioned concerning the presence of their fingerprints on looted valuables and shell casings at mass graves, roughly half of the prisoners decline to answer or are highly evasive. The other half admit to their participation in mass killings and looting but state they were acting under coercion. Their accounts of the times and locations of the atrocities accord with evidence obtained through exploiting captured documents. The cooperating prisoners identify several commanders as responsible for the atrocities using pre-conflict media footage to identify them. These commanders are also consistently identified by multiple members of the local civilian populace as being responsible for the mass killings and theft.

The pre-conflict photographs of the individuals suspected of ordering and leading the atrocities are electronically compared to the facial photographs of enemy prisoners of war under UN control. This results in two matches. When confronted both individuals admit they presented false identity documents upon capture purporting to be members of the medical service corps. They state they did this in order to protect their knowledge of their country's military and intelligence apparatus. When questioned regarding illegal killings and looting both individuals quickly claim no knowledge of such things and decline to answer further questions.

The investigation is pursued further. After considering its conclusions the UN Security Council refers the case the International Criminal Court. By the end of hostilities the International Criminal Court has issued several indictments for individuals who are later successfully prosecuted. The remaining prisoners of war are repatriated.

Tasks achieved using biometrics:

- Locate a person of interest
- Share identity information

B. REFERENCES

a. Privacy Act of 1974, 5 USC 552a, especially:

- 5 U.S.C. § 552a(b)(1)
- 5 U.S.C. § 552a(e)(9)
- J. 5 U.S.C. § 552a(e)(10)

b. Office of Management and Budget (OMB) Guidelines, 40 Federal. Register. 28,948, 28,955 (09 July 1975)

c. OMB CIRCULAR NO. A-130, Management of Federal Information Resources (08 February 1996)

d. Army Regulation 190-8, "Enemy Prisoners of War, Retained Personnel, Civilian Internees and Other Detainees," (October 1997), http://www.usapa.army.mil/pdffiles/r190_8.pdf

e. Title 10, United States Code (Armed Forces) (26 June 1998)

f. Federal Bureau of Investigation (FBI) "Electronic Fingerprint Transmission Specification" (EFTS) (January 1999)

g. FBI DNA Advisory Board, "Quality Assurance Standards for Forensic DNA Testing Laboratories and for Convicted Offender DNA Database Laboratories," (July 2000), http://www.fbi.gov/hq/lab/fsc/backissu/july2000/codispre.htm

h. Public Law 106-246 Military Appropriations (13 July 2000)

i. American National Standards Institute/National Institute of Standards and Technology (ANSI/NIST)-ITL 1- 2011, "Data Format for the Interchange of Fingerprint, Facial & Other Biometric Information"

j. Deputy Secretary of Defense Memorandum, "Executive Agent for the Department of Defense (DOD) Biometrics Project" (27 December 2000)

k. Assistant Secretary of Defense (Command, Control, Communication and Intelligence) (C3I) Memorandum "Biometrics as an Information Enabler" (19 January 2001)

l. A0380-19 Secretary of the Army Information Systems AIS "Change in Systems Records Notice" (amended 13 April 2001)

m. Deputy Secretary of Defense Memorandum " DOD Strategic Plan for Biometrics" (28 June 2002)

n. Army General Order No. 3 "Assignment of Functions and Responsibilities Within Headquarters, Dept. of the Army" (09 July 2002)

o. DOD Directive 5101.1 "DOD Executive Agent" (03 September 2002)

p. Secretary of the Army General Counsel Memorandum " Legislative Authority for the DOD Biometrics Program " (17 October 2002)

q. DOD Directive 8500.1 "Information Assurance (IA)" (24 October 2002)

r. DOD Instruction 8500.2 "Information Assurance Implementation" (06 February 2003)

s. DOD Directive 5000.01 "The Defense Acquisition System" (12 May 2003)

t. DOD Instruction 5000.02 "Operation of the Defense Acquisition System " (8 December 2008)

u. Deputy Secretary of Defense Memorandum " DOD Biometrics Enterprise Vision" (25 August 2003)

v. Homeland Security Presidential Directive HSPD-6 "Integration and Use of Screening Information" (16 September 2003), http://www.whitehouse.gov/news/releases/2003/09/20030916-5.html

w. Homeland Security Presidential Directive 7 "Critical Infrastructure Identification, Prioritization and Protection Purpose" (17 December 2003)

x. Assistant Secretary of Defense for Networks and Information Integration Memorandum, "Department of Defense Compliance with the Internationally Accepted Standard for Electronic Transmission and Storage of Fingerprint Data from 'Red Force' Personnel" (02 February 2004)

y. Army CIO/G-6 Memorandum "FBI Guidance on Collection of Fingerprint and Other Biometric Data from Military Detainees" (18 February 2004)

z. Deputy Secretary of Defense Memorandum, "Criteria and Guidelines for Screening and Processing Persons Detained by the Department of Defense in Connection with the War on Terrorism" (20 February 2004)

aa. Commander, Fleet Force Command Message "Establishment of a Navy Program to Support Counter Terrorism-Anti Terrorism-Force Protection" (12 April 2004)

bb. Turner Congressional Memorandum "DOD Collection and Use of Biometric Data" (12 April 2004)

cc. ANSI/INCITS 385-2004, "Face Recognition Format for Data Interchange" (May 2004). This standard is copyrighted and licensed copies are available from the International Organization for Standardization.

dd. DODD 8521.01E "Department of Defense Biometrics" (February 2008)

ee. ISO/IEC 19794-6:2005 379-2004, "Information technology -- Biometric data interchange formats -- Part 6: Iris image data". This standard is copyrighted and licensed copies are available from International Organization for Standardization ISO.

ff. US Central Command (CENTCOM) "Fragmentary Order Biometric Collection and Reporting System " (18 June 2004) gg. Title 28 United States Code Section 534, Acquisition, Preservation and Exchange of Identification Records and Information (19 June 2004)

hh. DOD Directive 1000.25, "Personnel Identity Protection" (19 July 2004)

ii. Assistant Secretary of Defense for Networks and Information Integration Memorandum ASD (NII), Director, Information Assurance Memorandum Establishment of a DOD Automated Biometric Identification System (ABIS) (05 August 2004)

jj. Deputy Secretary of Defense Memorandum, "DOD Detainee Biometric Collection Processing Policy" (15 August 2004)

kk. Biometrics Identity Management Agency (BIMA) "Enemy Prisoner of War Records Security Standard Operation Procedures" (23 August 2004)

ll. Homeland Security Presidential Directive 11 "Comprehensive Terrorist-Related Screening Procedures" (27 August 2004), http://www.whitehouse.gov/news/releases/2004/08/20040827-7.html or http://www.fas.org/irp/offdocs/direct.htm

mm. Homeland Security Presidential Directive 24

nn. Executive Order 13356, "Strengthening the Sharing of Terrorism Information to Protect America," (27 August 2004), http://www.whitehouse.gov/news/releases/2004/08/20040827-4.html

oo. Director, Biometrics Management Office (BMO) Memorandum " Interim Policy on Searching and Matching Biometric Data for the Iraqi Multi-Purpose Access Card" (20 October 2004)

pp. Deputy Secretary of Defense Memorandum, "Department of Defense Detainee Biometric Policy" (01 November 2004) (Classified)

qq. DOD Directive 5400.11, "DOD Privacy Program" (16 November 2004)

rr. Fleet Judge Advocate Memorandum "Legal Opinion on Legality of Taking Fingerprints of Suspected Terrorists During Conduct of Title 10 Missions" (08 December 2004)

ss. Director, Navy Staff Memorandum "Resource Sponsorship for Biometrics" (16 December 2004)

tt. National Security Presidential Directive 41/ Homeland Security Presidential Directive 13 "Maritime Security Policy" (21 December 2004)

uu. Chairman of the Joint Chiefs of Staff "National Military Strategy" (2011)

vv. Chief for Naval Operations Memorandum "Authority and Capabilities Required for the Global War on Terrorism" (18 January 2005)

ww. DOD Department of the Army Privacy Act of 1974, System of Records Notice (DOD ABIS) (25 February 2005)

xx. Secretary of Defense, "National Defense Strategy of the United States of America" (June 2008)

yy. DA-G6 Memorandum, "DOD Standard Operating Procedure (SOP) For Collecting and Processing Detainee Biometric Data" (04 March 2005)

zz. Defense Information Systems Agency (DISA) Memorandum "Interim Approval to Connect (IATC) for the Biometrics Identity Management Agency (BIMA), West Virginia to the Secret IP Router Network (SIPRNet)" (17 March 2005)

aaa. Deputy Secretary of Defense Memorandum, "Force Protection Identity Screening Policy for Base Access" (29 March 2005)

bbb. Deputy Chief of Staff for Intelligence Multi-National Force Iraq, Memorandum "CENTCOM Operational Needs Statement" (30 April 2005)

ccc. National Security Agency Information "Assurance Architecture, version 1.1" (May 2005)

ddd. DOD Identity Protection and Management Vision (June 2005)

eee. DOD "Electronic Biometric Transmission Specification, Version 1.1" (23 August 2005)

fff. Net-Centric Environment Joint Functional Concept (07 April 2006)

ggg. FBI Electronic Fingerprint Transmission Specification 7.1 (02 May 2005)

hhh. Department of The Army Office of The General Counsel memo, "Ownership of Biometric Data Submitted to DOD Automated Biometric Identification System (ABIS)" (12 May 2005)

iii. Deputy Secretary of Defense Memorandum "Joint Improvised Explosive Device (IED) Defeat" (27 June 2005)

jjj. Deputy Secretary of Defense Memorandum, "Department of Defense Policy for Biometric Identification System for Access to U.S. Installations and Facilities in Iraq" (15 July 2005)

kkk. DOD Biometrics Management Office Memorandum, "Interim Notification Process for Matches Made on Biometric Automated Toolset (BAT) Data using the DOD Automated Biometric Identification System (ABIS)" (15 July 2005)

lll. Capstone Concept for Joint Operations Version 2.0 (August 2005)

mmm. Deputy Secretary of Defense Memorandum "Notifying Individuals When Personal Information is Lost, Stolen, or Compromised" (15 July 2005)

nnn. Biometrics Identity Management Agency (BIMA) Memorandum, "Interim Executive Agent Guidance on Ownership of Biometric Data Submitted to and Stored in the DOD Automated Biometric Identification System (ABIS)" (15 July 2005)

ooo. Global Information Grid Information Assurance Initial Capabilities Document (06 March 2006) ppp. Net-Centric Environment Joint Functional Concept (07 April 2006)

qqq. Deputy Secretary of Defense Memorandum, "Collection of Biometric Data from Certain U.S. Persons in the United States Central Command (USCENTCOM) Area of Responsibility (AOR)" (24 May 2006)

rrr. "Products certified for compliance with the FBI's Integrated Automated Fingerprint Identification System image quality specifications," http://www.fbi.gov/hq/cjisd/iafis/cert.htm

sss. DA Form 2663-R, "Fingerprint Card," http://www.apd.army.mil/pub/eforms/pdf/a2663_r.pdf

ttt. DA Form 4137, "Evidence/Property Custody Document," http://www.apd.army.mil/pub/eforms/pdf/a4137.pdf

uuu. Department of Defense Biometrics Website, http://www.biometrics.DOD.mil

vvv. Secretary of Defense Strategic Planning Guidance, 2006-2011

www. Public Law 93-579, Disclosure of Social Security Number

xxx. Homeland Security Presidential Directive 59

yyy. DOD Identity Protection and Management Senior Coordinating Group Charter

zzz. DOD Roadmap to Identity Superiority

aaaa. CJCSI 6212.01E Interoperability and Supportability of Information Technology and National Security Systems

bbbb. Biometrically Enabled Intelligence Concept of Operations (26 Feb 10)

cccc. Joint Capabilities Document, Biometrics Support to Identity Management (31 Jan 08)

dddd. Deputy Secretary of Defense Memorandum "Authority to Collect, Store, and Share Biometric Information of Non-U.S. Persons with US Government Entities and Partner Nations" (13 Jan 12)

eeee. Biometric Glossary, www.biometrics.gov/ReferenceRoom/Introduction

ffff. DOD Directive 5205.15E "DOD Forensic Enterprise" (26 April 11)

gggg. DOD Directive 8521.01E "Department of Defense Biometrics" (21 February 08)

C. GLOSSARY

Analyze – To deliberately consider available information on an individual and reach logical conclusions. These conclusions can include his intent, affiliation(s), activities, location and behavioral patterns. (DOD Capstone Concept of Operations for Employing Biometrics in Military Operations)

Associated Information – Non-biometric information about a person. For example, a person's name, personal habits, age, current and past addresses, current and past employers, telephone number, email address, place of birth, family names, nationality, education level, group affiliations and history, including such characteristics as nationality, educational achievements, employer, security clearances, financial and credit history. (DOD Capstone Concept of Operations for Employing Biometrics in Military Operations)

Authentication – The process of establishing confidence in the truth of some claim. The claim could be any declarative statement for example: "This individual's name is 'Joseph K.' " or "This child is more than 5 feet tall." In biometrics, "authentication" is sometimes used as a generic synonym for verification. [National Science and Technology Council (NSTC) Subcommittee on Biometrics, 16 February 2006]

Authoritative Source – The primary DOD approved repository of biometric information on a biometric subject. The authoritative source provides a strategic capability for access to standardized, comprehensive and current biometric files within the DOD and for sharing of biometric files with joint, interagency and designated multinational partners. The DOD may designate authoritative sources for various populations in accordance with applicable law, policy and directives. (DODI 8521.bb)

Behavioral Biometric Characteristic – A biometric characteristic that is learned and acquired over time rather than one based primarily on biology. All biometric characteristics depend somewhat upon both behavioral and biological characteristic. Examples of biometric modalities for which behavioral characteristics may dominate include signature recognition and keystroke dynamics. (NSTC Subcommittee on Biometrics, 16 February 2006)

Biological Biometric Characteristic – A biometric characteristic based primarily on an anatomical or physiological characteristic, rather than a learned behavior. All biometric characteristics depend somewhat upon both behavioral and biological characteristic. Examples of biometric modalities for which biological characteristics may dominate include fingerprint and hand geometry. (NSTC Subcommittee on Biometrics, 16 February 2006)

Biometric Automated Toolset – Army (BAT-A) – A multimodal biometric system that collects, stores, and shares fingerprints, iris images, and facial photography. It is used to enroll, identify, and track persons of interest, build digital dossiers on individuals that can include attached digital images, reports, documents, and include a wide variety of reports to include biographic, contextual, relationship, and interrogation reports. BAT-A has an internal biometric signature search/match capability and can be configured into either a mobile or handheld configuration. (PM DOD Biometrics, TCM BF)

Biometrics Enabled Intelligence – Intelligence information associated with and or derived from biometrics data that matches a specific person or unknown identity to a place, activity, device, component, or weapon that supports terrorist/insurgent network and related pattern analysis, facilitates high value individual targeting, reveals movement patterns, and confirms claimed identity. (DoDD 8521.01E)

Biometric File – The standardized individual data set resulting from a collection action that consists of a biometric sample and contextual data. (DOD Capstone Concept of Operations for Employing Biometrics in Military Operations)

Biometric Identity – A distinct, non-refutable set of physical and behavioral characteristics. (DOD Capstone Concept of Operations for Employing Biometrics in Military Operations)

Biometric Samples – Information or computer data obtained from a biometric sensor device. Examples are images of a face or fingerprint. (NSTC Subcommittee on Biometrics)

Biometric – Measurable physical characteristic or personal behavior trait used to recognize the identity or verify the claimed identity of an individual. (JP 1-02)

Biometrics – The process of recognizing an individual based on measurable anatomical, physiological and behavioral characteristics. (JP 1-02)

Capabilities Based Assessment (CBA) – The CBA is the Joint Capabilities Integration and Development System analysis process that includes four phases: the functional area analysis, the functional needs analysis, the functional solution analysis and the post independent analysis. The results of the CBA are used to develop a joint capabilities document or initial capabilities document. (CJCSM 3170.01B, 11 May 2005)

Collect – Obtain biometric and related contextual data from an object, system, or individual with, or without, his knowledge. (DOD Capstone Concept of Operations for Employing Biometrics in Military Operations)

Contextual Data – Elements of biographical and situational information (who, what, when, where, how, why, etc.) that are associated with a collection event and permanently recorded as an integral component of the biometric file. (DOD Capstone Concept of Operations for Employing Biometrics in Military Operations)

Database – A collection of one or more computer files. For biometric systems, these files could consist of biometric sensor readings, templates, match results, related end user information, etc. (NSTC Subcommittee on Biometrics, 16 February 2006)

Decide/Act – Take action based on a biometric file's match results and analysis of associated information. (DOD Capstone Concept of Operations for Employing Biometrics in Military Operations)

Defense Enrollment Eligibility Reporting System (DEERS) and the Real-Time Automated Personnel Identification System (RAPIDS) – Operational programs in support of resources/benefits management, critical defense missions, the Uniformed Services Identification (ID) Card program and awareness regarding benefits to which Uniformed Services personnel and their family members are entitled. (Defense Manpower Data Center)

Enrollment – The process of collecting a biometric sample from an end user, converting it into a biometric reference and storing it in the biometric system's database for later comparison. (NSTC Subcommittee on Biometrics, 16 February 2006)

Family of Joint Future Concepts – Provides the conceptual basis for capabilities-based assessments (CBAs) to answer these questions by identifying capabilities, gaps and redundancies as well as potential non-materiel and materiel approaches to addressing the issues. (CJCSM 3170.01B, 11 May 2005)

Forensic – Scientific analysis linking persons, places, things and events. These linkages are made in both traditional law-enforcement and medical purviews, as well as the expeditionary environment. (DODD 5205.15E, 26 April 2011)

Functional Capabilities Board – A permanently established body that is responsible for the organization, analysis and prioritization of joint warfighting capabilities within an assigned functional area. (CJCSM 3170.01B, 11 May 2005)

Future Joint Force – A force that is knowledge-empowered, networked, interoperable, expeditionary, adaptable / tailorable, enduring / persistent, precise, fast, resilient, agile and lethal. (CJCSM 3170.01B, 11 May 2005)

Homeland Security Presidential Directive – 12 – A policy for a common identification standard for federal employees and contractors. (DOD Capstone Concept of Operations for Employing Biometrics in Military Operations)

Identification – A task where the biometric system searches a database for a reference matching a submitted biometric sample and if found, returns a corresponding identity. A biometric is collected and compared to all references in a database. Identification is "closed-set" if the person is known to exist in the database. In "open-set" identification, sometimes referred to as a "watchlist," the person is not guaranteed to exist in the database. The system must determine whether the person is in the database, then return the identity. (NSTC Subcommittee on Biometrics and Identity Management Biometrics Glossary)

Identity – The set of attribute values (i.e. characteristics) by which an entity is recognizable and that, within the scope of an identity manager's responsibility, is sufficient to distinguish that entity from any other entity and to distinguish the identity from any other identity. (DOD Capstone Concept of Operations for Employing Biometrics in Military Operations)

Identity Management – The combination of technical systems, policies and processes that create, define, govern and synchronize the ownership, utilization and safeguarding of identity information. (DOD Identity Management Strategic Plan)

Identity Protection – The process of safeguarding and ensuring that identities of individuals, devices, applications and services are not compromised. (DOD IPMSCG Charter)

Individual – A specific, physical person. (DOD Capstone Concept of Operations for Employing Biometrics in Military Operations)

Information Assurance – Operational capabilities that facilitate information sharing while protecting and defending electronic information and information systems by ensuring their availability, integrity, authentication and confidentiality. (Pentagon Area Common Information Technology Wireless Security Policy, September 2002)

Intelligence Exploitation – The process of converting collected information into forms suitable to the production of intelligence. (Joint and National Intelligence Support to Military Operations JP 2-01)

Joint Capabilities Document (JCD) – The JCD identifies a set of capabilities that support a defined mission area as identified in the Family of Joint Future Concepts, concept of operations (CONOP), or Unified Command Plan-assigned missions. The capabilities are identified by analyzing what is required across all functional areas to accomplish the mission. The gaps or redundancies are then identified by comparing the capability needs to the capabilities provided by existing or planned systems. The JCD will be used as a baseline for one or more initial capabilities documents or joint doctrine, organization, training, materiel, leadership and education, personnel and facilities change recommendations, but cannot be used for the development of capability development or capability production documents. The JCD will be updated as changes are made to the Family of Joint Future Concepts, CONOP or assigned missions. (CJCSM 3170.01B, 11 May 2005)

Joint Force – A general term applied to a force comprised of significant elements, assigned or attached, of two or more Military Departments operating under a single Joint Force commander. (CJCSM 3170.01B, 11 May 2005)

Latent – A substance lying dormant or hidden until circumstances are suitable for development or manifestation. (Oxford Dictionary)

Local-Trusted Source – A sub-set of the authoritative source that is established to accomplish a specific function within an operational mission or business function. Reasons for establishing a local trusted source include: insufficient network connectivity able to provide immediate access to the authoritative source or an operational need for closed-loop access or permission application. If a match is not made against a local trusted source, then the file should be queried against the authoritative source. (DOD Capstone Concept of Operations for Employing Biometrics in Military Operations)

Local Un-trusted Source – A local repository of biometric files that have not been enrolled with an authoritative or local trusted source. In many cases, local un-trusted sources are established for missions of short duration or to satisfy political, policy, or legal restrictions related to the sharing of biometric information. (DOD Capstone Concept of Operations for Employing Biometrics in Military Operations)

Logical Access – Process of granting access to information system resources to authorized users, programs, processes, or other systems. The controls and protection mechanisms that limit users' access to information and restrict their forms of access to only what is appropriate. (DOD Capstone Concept of Operations for Employing Biometrics in Military Operations)

Match – A decision that a biometric sample and a stored template comes from the same human source, based their high level of similarity. (NSTC Subcommittee on Biometrics and Identity Management Biometrics Glossary)

Modality – A type or class of biometric system. For example: face recognition, fingerprint recognition, iris recognition, etc. (NSTC Subcommittee on Biometrics, 16 February 2006)

Multimodal Biometric System – A biometric system in which two or more of the modality components (biometric characteristic, sensor type or feature extraction algorithm) occurs in multiple. (NSTC Subcommittee on Biometrics, 16 February 2006)

Non-DOD Partners – Interagency and multinational partners. (DOD Capstone Concept of Operations for Employing Biometrics in Military Operations)

Normalize – Create a standardized, high-quality biometric file consisting of a biometric sample and contextual data. (DOD Capstone Concept of Operations for Employing Biometrics in Military Operations)

One-to-many – A phrase used in the biometrics community to describe a system that compares one reference to many enrolled references to make a decision. The phrase typically refers to the identification of an individual. (NSTC Subcommittee on Biometrics, 16 February 2006)

One-to-one – A phrase used in the biometrics community to describe a system that compares one reference to one enrolled reference to make a decision. The phrase typically refers to the verification task (though not all verification tasks are truly one-to-one). The identification task can be accomplished by a series of one-to-one comparisons. (NSTC Subcommittee on Biometrics, 16 February 2006)

Person of Interest – An individual whose identity is of special interest. (DOD Capstone Concept of Operations for Employing Biometrics in Military Operations)

Physical Access – The process of granting access to installations and facilities. (DOD Capstone Concept of Operations for Employing Biometrics in Military Operations)

Provide – Exchange analysis and associated information on individuals among approved DOD, interagency and international partners in accordance with applicable laws, policies, authorities and agreements. (DOD Capstone Concept of Operations for Employing Biometrics in Military Operations)

Share – Exchange standardized biometric files and match results among approved DOD, interagency and international partners in accordance with applicable laws, policies, authorities and agreements.

Source – An approved database and infrastructure that stores biometrics files. (DOD Capstone Concept of Operations for Employing Biometrics in Military Operations)

Store – Maintain biometric files to make available standardized, current biometric information of individuals when and where required. Biometric files are initially enrolled and then subsequently updated as part of storing. (DOD Capstone Concept of Operations for Employing Biometrics in Military Operations)

Verification – A task where a biometric system attempts to confirm an individual's claimed identity by comparing a submitted sample to one or more previously enrolled templates. (NSTC Subcommittee on Biometrics and Identity Management Biometrics Glossary)

D. ACRONYMS

ANSI - American National Standards Institute

AOR - Area of Responsibility

BAT - Biometric Automated Toolset

BEI - Biometrically Enabled Intelligence

BIMA - Biometrics Identity Management Agency

BISA - Biometrics Identification System for Access

CBA - Capabilities Based Assessment

CBP - Customs and Border Protection

CCJO - Capstone Concept for Joint Operations

CJIS - Criminal Justice Information Systems

DEERS - Defense Enrollment Eligibility Reporting System

DHS - Department of Homeland Security

DISA - Defense Information Systems Agency

DMDC - Defense Manpower Data Center

DOD - Department of Defense

DODD - Department of Defense Directive

DOTMLPF - Doctrine, Organization, Training, Materiel, Leadership & Education, Personnel, Facilities

EFTS - Electronic Fingerprint Transmission Specification

EOD - Explosive Ordnance Disposal

FBI - Federal Bureau of Investigation

FCB - Functional Capabilities Board

GIG - Global Information Grid

HSPD-12 - Homeland Security Presidential Directive - 12

IA - Information Assurance

IATC - Interim Approval to Connect

ICD - Initial Capabilities Document

IED - Improvised Explosive Device

IPMSCG - Identity Protection and Management Senior Coordinating Group

JCD - Joint Capabilities Document

JCIDS - Joint Capabilities Integration and Development Systems

KM/DS - Knowledge Management and Decision Support

KST - Known or Suspected Terrorist

LEDET - Law Enforcement Detachment

MEDCAPs - Medical Civic Action Programs

NCTC - National Counter-terrorism Center

NIST - National Institute of Standards and Technology NSTC - National Science and Technology Council

NGO - Non-Governmental Organization

PKI - Public Key Infrastructure

RAPIDS - Real-Time Automated Personnel Identification System

SIPRNet - Secret Internet Protocol Router Network

SMT - Scar Mark & Tattoo

SOF - Special Operations Forces

SOP - Standard Operation Procedures

TSC - Terrorist Screening Center

USCENTCOM - United States Central Command

www.ingramcontent.com/pod-product-compliance
Lightning Source LLC
Chambersburg PA
CBHW080620180526
45168CB00007B/2987